BLOODTHIRSTY PLANTS
AN IMAGINATION LIBRARY SERIES

BLADDERWORTS
Trapdoors to Oblivion

By Victor Gentle

With special thanks to the people at
the Carolina Biological Supply Company,
and to Sebastian Vieira,
for their kind encouragement and help.

Gareth Stevens Publishing
MILWAUKEE

For a free color catalog describing Gareth Stevens' list of high-quality books and multimedia programs, call 1-800-542-2595 (USA) or 1-800-461-9120 (Canada).
Gareth Stevens Publishing's Fax: (414) 225-0377.
See our catalog, too, on the World Wide Web: http://gsinc.com

Library of Congress Cataloging-in-Publication Data

Gentle, Victor.
　Bladderworts: trapdoors to oblivion / by Victor Gentle.
　　p. cm. — (Bloodthirsty plants)
　Includes bibliographical references (p. 23) and index.
　Summary: Introduces these carnivorous plants by describing their traps, plant types, diet, habitat, and possible use against the cane toad plague in Australia.
　ISBN 0-8368-1654-4 (lib. bdg.)
　1. Bladderworts–Juvenile literature.　[1. Bladderworts.　2. Carnivorous plants.]　I. Title.
II. Series: Gentle, Victor.　Bloodthirsty plants.
QK495.L53G46　1996
583'81–dc20　　　　　　　　　　　　　　　　　　　　　　　　　　96-5614

First published in 1996 by
Gareth Stevens Publishing
1555 North RiverCenter Drive, Suite 201
Milwaukee, WI 53212 USA

Text: Victor Gentle
Page layout: Victor Gentle and Karen Knutson
Cover design: Karen Knutson
Photo credits: Cover (main), pp. 5, 7, 9 © Visuals Unlimited/Cabisco; cover (background) © Stuart Wasserman/Picture Perfect; p. 11 © Visuals Unlimited/Prance; p. 13 © Visuals Unlimited/Dick Poe; p. 15 © Sebastian Vieira; p. 17 © Visuals Unlimited/Kirtley-Perkins; p. 19 © Visuals Unlimited/Joe McDonald; p. 21 © David M. Stone 1981/PHOTO/NATS

© 1996 by Gareth Stevens, Inc. All rights reserved to Gareth Stevens, Inc. No part of this book may be reproduced, stored in a retrieval system, or transmitted in any form or by any means, electronic, mechanical, photocopying, recording, or otherwise, without the prior written permission of the publisher except for the inclusion of brief acknowledged quotations in a review.

Printed in the United States of America

1 2 3 4 5 6 7 8 9 01 00 99 98 97 96

TABLE OF CONTENTS

A Water Flea Is Caught 4
The Vacuum Trap 6
Digesting Its Meal 8
Survival of the Best Fed 10
Bladderworts on Soggy,
   Wet Land! 12
Bladderworts — Up a Tree! 14
Bladderworts — In the Water! 16
Helping to Fight Cane Toads? 18
Sundry Schemes to Snare
   a Snack 20
Growing Bladderworts Yourself . . . 22
Where to Get Plants or Seeds 22
More to Read and View 23
Where to Write to Find Out More . . 23
Glossary 24
Index 24

A WATER FLEA IS CAUGHT

A small water flea swims in and out of underwater leaves and stems. It darts to investigate a small, flat, disklike bag. Accidentally, the water flea touches one or two stiff hairs. A tiny trap door whips open. The water flea is sucked inside, and the trap door springs shut.

The unlucky prisoner is about to become a meal for a **carnivorous** plant — a plant that eats animals! This "bloodthirsty" plant is one of the 200 to 300 **species** of the **genus** *Utricularia* (yoo-TRIK-yoo-LAH-ryuh) — the bladderworts.

Other carnivorous plants have their own ways of catching food. Bladderworts have developed a deadly, high-speed vacuum trap!

A tiny, water-living insect trapped inside a bladderwort vacuum trap, shown here about 30 times bigger than life size.

THE VACUUM TRAP

The traps, called **bladders**, look like flat, empty bags. The bladder walls strain to spring open, but they cannot. The trapdoor is held shut by a thin film of glue that seals the entrance. As a result, there is a partial vacuum inside the bladder. When a small animal presses the trip hairs, the seal is broken. It takes less than one-fiftieth of a second for the trapdoor to flip open and the victim to be sucked in.

Bladderworts have the smallest traps of all green carnivorous plants. The biggest are just one-fifth of an inch (5 millimeters) long. The smallest are a tiny one-hundredth of an inch (0.25 mm) long.

Only the carnivorous **fungi** have smaller traps.

Some of the largest traps can be found on the greater bladderwort (*Utricularia vulgaris*), shown here about 9 times bigger than life size.

DIGESTING ITS MEAL

As soon as a meal has been caught, the bladder starts to remove some of the water. At the same time, special **glands** release **acids** and **enzymes**. These **dissolve** all the soft parts of the animal. The **nutrients** in this "soup" are absorbed by these same glands to feed the plant. Bit by bit, the bladder is emptied. A partial vacuum is made once more. The trap is ready for action again.

Underwater traps on a bladderwort plant in the process of digesting meals, shown here about 25 times bigger than life size.

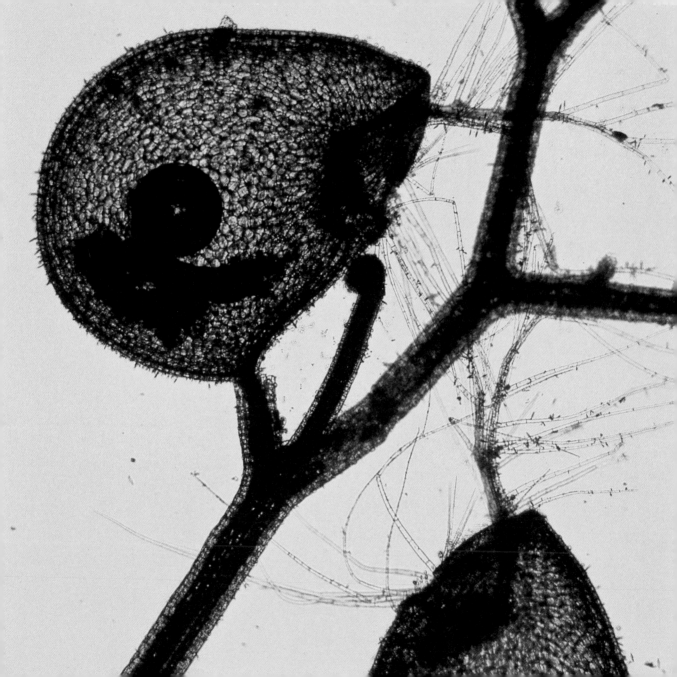

SURVIVAL OF THE . . . BEST FED

Bladderworts live in areas where **minerals** that plants need are in short supply. This is true of many swampy, watery habitats. These are the favorite habitats of most carnivorous plants. Here, carnivorous plants have an advantage. By eating insects and other tiny animals, they find the essential nutrients they need. This helps them become strong. It also helps them as they compete with other nearby plants for living space.

The yellow flowers of an aquatic bladderwort reach up from the water's surface. The traps are all under water, busily feeding.

BLADDERWORTS ON SOGGY, WET LAND!

Bladderworts can be found in the wild in almost every country of the world.

Most bladderworts grow on wet, boggy soil, on wet moss, or in other wet areas. The tiny traps are usually hidden beneath the surface. There they wait to catch small animals that live in the watery soil.

All bladderworts produce small, beautiful flowers. Depending on the species, these delicate flowers may be pink, white, yellow, blue, or red. There may be just one flower, or up to about thirty, on a single leafless stem.

A good habitat for bladderworts — a pine barrens bog in New Jersey, in the United States. The yellow flowers are horned bladderworts, *Utricularia cornuta* (kor-NEW-tuh).

BLADDERWORTS... UP A TREE!

In the rain forests of South America, some species of bladderworts live on trees. They live on the rotting bark, rotting leaves, and moss in the nooks and crannies of trunks and branches.

Because they do not take nutrients from the living tree itself, they are called **epiphytes**. (Some plants do take nutrients from the plants they grow on. They are called **parasites**). Some epiphytic bladderworts also live in the rain forests of Central America and some Caribbean islands.

A tree-dwelling bladderwort, *Utricularia jamesoniana* (jame-SO-nee-ANN-uh), in Colombia, South America.

BLADDERWORTS... IN THE WATER!

Many bladderworts live in water — in ponds, lakes, and wetland areas. They trap water fleas, worms, and the small larvae of insects like mosquitoes. Some even catch tiny fish.

One interesting species, the swollen bladderwort *Utricularia inflata* (inn-FLAH-tuh), grows special bladders. The bladders are filled with air. They allow the plant to float like a raft. The swollen bladderwort's leaves and traps hang beneath the surface of the water.

The swollen bladderwort, *Utricularia inflata*. This unusual plant also has special air bladders that help it float at the water's surface.

HELPING TO FIGHT CANE TOADS?

In 1935, two crates of giant tropical toads arrived in Australia from Hawaii. The toads had a fancy scientific name, but most Australians quickly came to know them as cane toads. Two species of beetles were killing Australia's sugarcane crops. The toads were brought in to fight the pests.

Big mistake. The cane toads showed little interest in eating the beetles. Worse still, the toads had few natural enemies in Australia. They reproduced quickly and were soon out of control. Cane toads became a bigger pest than the beetles ever were.

Now there's a rumor going around that some bladderworts like to dine on cane toad tadpoles . . . . But even if that's true, it's unlikely that the bladderworts could eat fast enough!

Excited government officials called them the "Introduced Tropical American Toads, *Bufo marinus*." The animals quickly became known as "cane toads."

SUNDRY SCHEMES TO SNARE A SNACK

The bladderworts have the smallest traps of all green carnivorous plants. They are the only ones to suck in their prey. Other bloodthirsty plants, with bigger traps, use different methods.

Sundews trap insects and small animals with a sweet, sticky liquid. Butterworts trap their **prey** on greasy leaves. Other plants (the Venus fly trap, for one) have highly sensitive spring traps. Yet others, called pitcher plants, use a slippery pit of no escape. The largest pitcher plants can eat animals as big as small monkeys. Some hungry fungi even lasso their prey!

You can learn more about the strange and wonderful world of carnivorous plants by reading other books. And you can grow your own!

The pitcher trap of a northern pitcher plant, *Sarracenia purpurea* (SAR-ra-SEE-nyuh per-per-REE-yuh) stands about 3 inches (7.5 cm) tall behind a mass of sticky sundew plants.

GROWING BLADDERWORTS YOURSELF

Bladderworts that live in water can be grown in aquariums quite easily. Bladderworts that grow in boggy soils can be grown in sphagnum moss or peaty soil inside large containers with **humid** air.

It's best to get specific instructions about the plants from the people who supply them.

WHERE TO GET PLANTS OR SEEDS

Here are some addresses of carnivorous plant suppliers. For other sources, contact a club or society listed on the next page.

Heldon Nurseries
Ashbourne Road
Spath Uttoxeter, ST14 5AD
England

Carolina Biological Supply Company
2700 York Road
Burlington, NC 27215
USA

Peter Paul's Nursery
4665 Chapin Road
Canandaigua, NY 14424
USA

Exotica Plants
Community Mail Bag
Cordalba, QLD 4660
Australia

Hillier Water Gardens
Box 662, Qualicum Beach
BC V9K 1T2
Canada

Silverhill Seeds
P.O. Box 53108, Kenilworth 7745
Republic of South Africa

MORE TO READ AND VIEW

Books (nonfiction):
- *Butterworts: Greasy Cups of Death.* Victor Gentle (Gareth Stevens)
- *Carnivorous Mushrooms: Lassoing Their Prey?* Victor Gentle (Gareth Stevens)
- *Carnivorous Plants.* Nancy J. Nielsen (Franklin Watts)
- *Killer Plants.* Mycol Doyle (Lowell House Juvenile)
- *Pitcher Plants: The Elegant Insect Traps.* Carol Lerner (Morrow)
- *Pitcher Plants: Slippery Pits of No Escape.* Victor Gentle (Gareth Stevens)
- *Plants of Prey.* Densey Clyne (Gareth Stevens)
- *Sundews: A Sweet and Sticky Death.* Victor Gentle (Gareth Stevens)
- *Venus Fly Traps and Waterwheels.* Victor Gentle (Gareth Stevens)

Books (fiction):
- *Elizabite: Adventures of a Carnivorous Plant.* H.A. Rey (Linnet)
- *Island of Doom.* Richard Brightfield (Gareth Stevens)

Videos (nonfiction): *Carnivorous Plants.* (Oxford Scientific Films)

Videos (fiction): *The Day of the Triffids* and *The Little Shop of Horrors* are fun to watch.

WHERE TO WRITE TO FIND OUT MORE

Your community may have a local chapter of a carnivorous plant society. Try looking it up in the telephone directory. Or contact one of the following national organizations:

Australia
Australian Carnivorous Plant Society, Inc.
P.O. Box 391
St. Agnes, South Australia 5097 Australia

New Zealand
New Zealand Carnivorous Plant Society
P.O. Box 21-381, Henderson
Auckland, New Zealand

United Kingdom
The Carnivorous Plant Society
174 Baldwins Lane, Croxley Green
Hertfordshire WD3 3LQ
England

Canada
Eastern Carnivorous Plant Society
Dionaea, 23 Cherryhill Drive
Grimsby, Ontario, Canada L3M 3B3

South Africa – has no CP society, but a supplier to contact is:
Eric Green, 11 Wepener Street
Southfield, 7800, Cape, South Africa

United States
International Carnivorous Plant Society
Fullerton Arboretum
California State University at Fullerton
Fullerton, CA 92634 USA

If you are on the Internet, or otherwise on-line, you can call up a World Wide Web page that gives links to other Web pages of interest to carnivorous plant enthusiasts: http://www.cvp.com/feedme/links.html

GLOSSARY

You can find these words on the pages listed. Reading a word in a sentence helps you understand it even better.

acids (ASS-idz) — harsh liquids that can dissolve many things 8

bladders (BLADD-erz) — baglike parts of plants or animals 6, 8, 16

carnivorous (kar-NIV-er-us) — flesh-eating 4, 6, 10, 20

dissolve (dih-ZOLV) — to make into a solution 8

enzymes (EN-zimes) — special substances that help digestion 8

epiphytes (EP-uh-fites) — plants that grow on other plants without stealing nutrients from them 14

fungi (FUN-JYE) — plural: **fungus** — a type of plant without the special green substance found in most leafy plants, so it can't make its own food from sunlight, air, and water 6, 20

genus (JEE-nus) — group of closely related plants or animals 4

glands (GLANZ) — special plant parts that produce and absorb liquid substances 8

humid (HYOU-mid) — damp 22

minerals (MIN-uh-rullz) — non-living materials, such as earth, rock, or salt 10

nutrients (NOO-tree-unts) — substances with good food value 8, 10, 14

parasites (PARE-uh-sites) — plants or animals that live off other plants or animals and steal nutrients from them 14

prey (PRAY) — a victim of a hunter, trapper, or trap 20

species (SPEE-shees) — an individual type of plant or animal 4, 12, 16, 18

INDEX

aquariums 22
Australia 18
bark 14
bladders 6, 8, 16
bloodthirsty 4, 20
bogs 12
Bufo marinus 18
butterworts 20
cane toads 18
Caribbean islands 14

Central America 14
flowers 10, 12
greater bladderworts 6
habitats 10, 12
leaves 4, 14, 16, 20
mosquito larvae 16
moss 12, 14, 22
pitcher plants 20
prisoners 4

rain forests 14
Sarracenia purpurea 20
South America 14
stems 4, 12
sugarcane 18
sundews 20
trapdoors 4, 6
trip hairs 6
Utricularia cornuta 12
Utricularia inflata 16

Utricularia jamesoniana 14
Utricularia vulgaris 6
vacuum traps 4
Venus fly traps 20
victims 6
water fleas 4, 16
worms 16

Poplar Grove K-4 Media Center